Ebenezer Baldwin Andrews

Letter Of Prof E.B. Andrews

On the Coal and Iron Deposits of the Upper Sunday Creek and Moxahala

Valleys in Perry County, Ohio

Ebenezer Baldwin Andrews

Letter Of Prof E.B. Andrews
On the Coal and Iron Deposits of the Upper Sunday Creek and Moxahala Valleys in Perry County, Ohio

ISBN/EAN: 9783337047023

Printed in Europe, USA, Canada, Australia, Japan

Cover: Foto ©berggeist007 / pixelio.de

More available books at **www.hansebooks.com**

LETTER

OF

PROF. E. B. ANDREWS,

OF THE OHIO GEOLOGICAL CORPS,

ON THE

COAL AND IRON DEPOSITS

OF THE

UPPER SUNDAY CREEK AND MOXAHALA VALLEYS,

IN

PERRY COUNTY, OHIO.

COLUMBUS:
PRINTED AT OHIO STATE JOURNAL OFFICE.

1873.

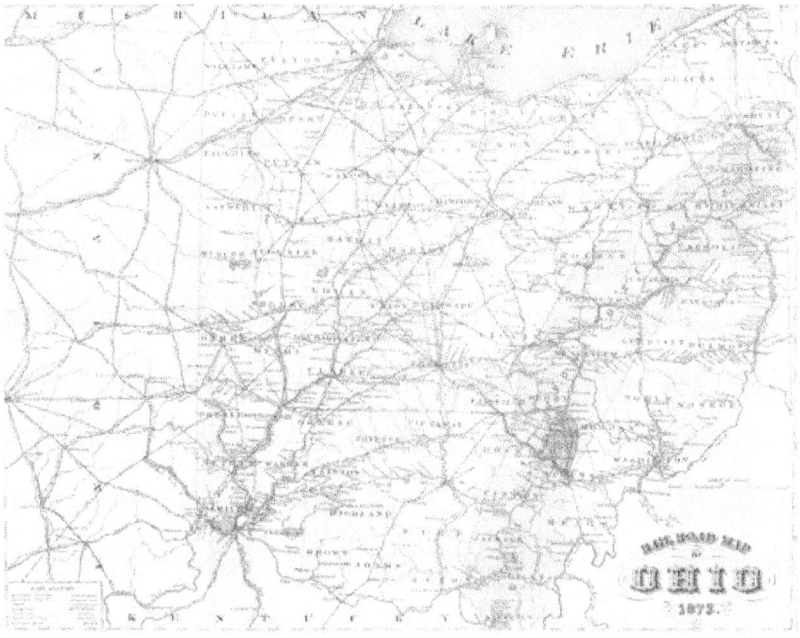

RAIL ROAD MAP OF OHIO 1873.

PROF. E. B. ANDREWS

ON THE

COAL AND IRON DEPOSITS IN PERRY COUNTY, OHIO.

HON. THOMAS EWING,

LANCASTER, OHIO,

SIR:—At your request, I have made a careful examination of the mineral lands, owned by several Mining Companies, in Monroe, Salt Lick, Pleasant and Pike townships, in the southern part of Perry County, Ohio, on the upper waters of Sunday creek, and on the South fork of the Moxahala. Perry County has long been known to contain very valuable seams of coal, but without suitable railroads the vast resources could not be turned to any profit. Such railroads are now built or are in process of construction, and this finest coal-field west of the Allegheny Mountains is arresting no little attention. At various points in the Monday creek valley the coal is already largely mined, and preparations are being made for practical mining operations of equal and perhaps greater magnitude in the upper Sunday creek valley. It is in the latter valley that the coal estates of these companies are chiefly situated, embracing many thousands of acres, wisely selected, so as to include all, or very nearly all the best coal lands.

It is here that I found the maximum thickness of coal of the State, and indeed, as I believe, in all the West. The quality is very superior, and shows an adaptation of the coal to a large number of uses.

The *general situation* of these properties is most advantageous. They are on the western margin of the great Allegheny coal-field and consequently border the vast coalless district, which extends over **two-thirds** of Ohio, **and over all of** Indiana, except its western **and** south-western border. Northern Illinois, Wisconsin, Michigan and Western Canada are also largely dependent **on** coal obtained in Ohio. This coalless district contains many large cities, such **as** Cincinnati, Chicago, Cleveland, Dayton, Toledo, &c., &c. The coal-field in Perry County is so centrally located that it may command the markets over a very wide range of country. It can supply largely Cleveland, Toledo and other Lake cities, and also to a considerable extent Cincinnati and other more southern points **of importance. This coal** will, therefore, **radiate, so to** speak, **to markets in all** directions. Being on the border **of** the great level country which stretches **far** away to the West and Northwest, the railroads will have relatively light grades, and can carry heavy freights at the least expense. These general advantages of location I regard as very great. Besides large shipments of coal, much will be profitably consumed in manufactories at or near the points of production, and I look forward to the time when there will **be** very large manufacturing towns and cities in the neighborhood of the great mines of Perry and adjacent counties.

All classes of manufactures in which cheap fuel is the **most important** element will **be drawn** toward this coal-field. Iron manufactures needing coal **of** special excellence and purity **will** find this region **an** attractive **one.** There is in the district a large amount **of good** iron ore, and more can be cheaply obtained from the Northern Lakes, brought as return freight by the cars taking the coal to Cleveland, Toledo, Chicago, &c.

With these general introductory statements, I proceed to give **you a** detailed account of the mineral resources found on the **property of** the several companies.

COAL.

THE GREAT SEAM.

To prevent confusion, I propose to follow the fortunes of this principal seam of coal, and note its development and structure at the several points where exposures **are** found on these estates.

Beginning on the extreme western limit on the West branch of the West fork of Sunday creek, high up the stream on the Samuel Turner place, south-west quarter, section 11, Salt Lick township, we find the first exposure of coal. The seam is divided by thin partings into three benches of 2 feet 1 inch, 11 inches, and 2 feet 6 inches respectively, beginning at the top, making a total of 5 feet 6 inches of coal. The quality of the coal is undoubtedly good. Over it are 10 feet of slate, and nodules of siderite ore were seen in the clay shales about 10 feet below. It is quite probable that the coal increases in thickness rapidly to the east, for on other branches of the West fork, in a line nearly due east, it is found in its maximum development.

The Lyons bank, on West fork, below the Turner land, shows a thickness of coal of 7 to 8 feet. Below, in the vicinity of the Sulphur Spring, the coal is a little over 6 feet thick. Below this point on this stream, the seam has been subjected to the abrading action of ancient currents of water, and is much of it often removed and replaced by sand rock brought in, in the form of sand, by the same currents.

At an old opening into the coal, by the roadside, west of the residence of Benjamin Sanders, in the south-east quarter of section 24, Salt Lick township, we find the top of the coal removed and the sandrock usurping its place. A section at this point is as follows:

(BENJAMIN SANDERS,)

Fig. 1.

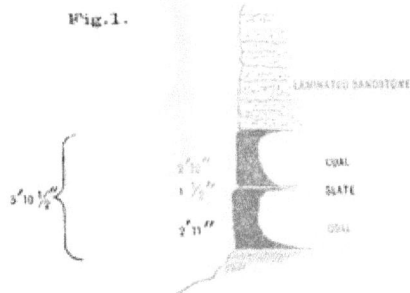

LAMINATED SANDSTONE

3′ 10 ½″

2′ 1 ½″   COAL

1 ½″   SLATE

2′ 11″   COAL

By this section it will be seen that the upper and more than half of the middle benches of coal are gone. The middle bench, when fully developed, measures 5 feet 9 inches, and about 40 rods, by estimate, east of the exposure last mentioned, the upper bench is by measurement 3 feet 4 inches. The upper bench, however, varies much in thickness in different parts of these lands.

In the immediate bank of the stream nearly south of the house of Benjamin Sanders there is a partial exposure of the seam, but only of the upper part. Here the coal is evidently very thick. A little east of this exposure we find a local interference with the continuity of the seam. There was, evidently, after its formation, a channel way cut through the coal, and this has been filled with clay. I believe this break in the seam to be only local and limited. Similar difficulties are not uncommon in seams of coal.

Further down the West fork, we find, on the south side of the stream, near the middle of section 19, Monroe township, the old Benjamin Sanders bank. Here the coal seam presents the following structure:

BENJAMIN SANDERS.— S. OF W. FORK,

Fig. 2.

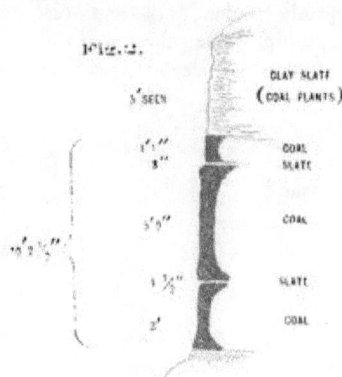

The upper bench is less thick than usual, but I think this is only a local matter, for farther up the stream, opposite Mr. Sanders' house, the upper bench measures 3 feet 4 inches, and at the Cam. Sanders bank, a little below, it measures 3 feet. The coal of the Benjamin Sanders bank is of very superior quality and can not fail to give satisfaction in any market. A sample of the coal from the middle bench of this bank was taken by myself in 1869, and placed in the hands of Prof. Wormley, the chemist of the Ohio Geological Survey. The following are the results of his analysis :

Specific gravity,	1.300
Water,	5.60
Ash,	2.03
Volatile matter,	29.92
Fixed carbon,	62.45
Total,	100.00
Sulphur,	0.76

This analysis indicates a remarkably pure and valuable coal. The percentage of ash is unusually small, and that of the sulphur is also small. Should no more than half of the sulphur pass off in coking, (and generally the coal of this seam loses much more than half in coking), this coal will serve an excellent purpose for the blast furnace. The very large percentage of fixed carbon will give the coal great heating power. No sample of coal from southern Perry County has afforded an analysis more satisfactory in all respects. If this represents proximately the quality of the coal to be found in the hills lying south of this part of West fork, this portion of your lands is destined to be immensely valuable. The better part of the seam for furnace use will perhaps be the middle bench, which is, in its upper 12 inches, and lower 18 inches, very rich in mineral charcoal, and shows an unusual degree of lamination.

On the north side of the stream an opening has been made recently into the coal at the base of the hill on the land of

A. Sanders, on the north-east quarter of section 19, Monroe township, and the coal presents the following divisions :

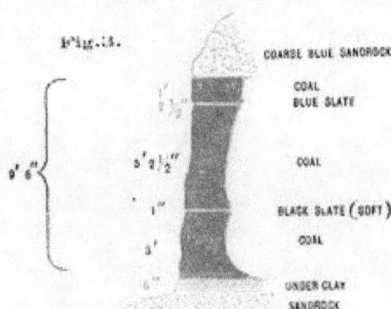

(A. SANDERS,)

Here it appears that there is a local interference from sandrock, which has taken the place of the usual roof slate and of a small portion of the coal. That this is only a local intrusion of the sandrock is evident from the fact that at all the exposures nearest this location the shales are found above the coal. The quality of the coal is evidently very superior. The lower 8 inches of the middle bench is highly laminated with thin films of sedimentary partings. This will not injure the coal except to increase somewhat the percentage of ash.

The entry appears to be well made and thoroughly timbered. The roof of the coal, whether the local and exceptional sandrock or the more usual slates, will, I think, be firm and all that could be desired.

The following analyses have been made of two samples of coal from the bank on the A. Sanders place :

No. 1, taken from the middle of the middle bench (5 feet 2½ inches thick.)
No. 2, taken from the middle of the lower bench (3 feet thick.)
No. 3, average of the two, representing 8 feet 2½ inches coal.

	No. 1.	No. 2.	No. 3.
Specific gravity,	1.315	1.328	1.321
Moisture,	5.20	4.70	4.95
Ash,	3.50	7.00	5.25
Volatile combustible matter,	30.80	31.30	31.05
Fixed carbon,	60.50	57.00	58.75
Total,	100.00	100.00	100.00
Sulphur,	0.68	1.01	0.845
Sulphur left in coke,	0.41	0.68	0.545
Gas per lb. in cubic feet,	3.64	3.56	3.60
Ash,	Yellow.	Gray.	

The average of the two analyses, given in No. 3, shows a very superior coal.

The Sanders coal gives 0.42 per cent. of a cubic foot more gas to the lb. of coal than the New Straitsville coal, which is justly popular as a gas-coal, from the high illuminating power of the gas. As yet no trials have been made of the coal of the Sunday creek valley in this particular. I know of no reason why there should be any difference. Both coals belong to the same continuous seam, and both present, to a very considerable degree, the same physical appearance and a general similarity of result in chemical analysis.

On the Cam. Sanders farm, a little east of the new opening last described, I find an old opening where the coal seam has a fine development, a section of which is here given.

(C. SANDERS' BANK,)

Fig. 1.

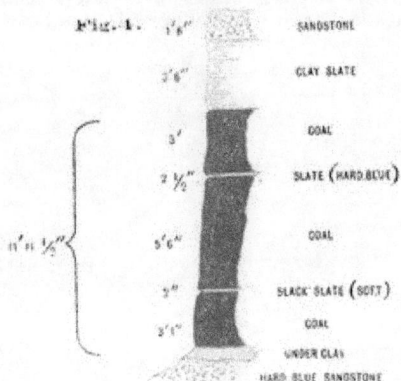

1' 6"		SANDSTONE
2' 6"		CLAY SLATE
3'		COAL
2 ½"		SLATE (HARD BLUE)
5' 6"		COAL
2"		BLACK SLATE (SOFT)
3' 1"		COAL
		UNDER CLAY
		HARD BLUE SANDSTONE

11' 11 ½"

Here the seam shows the two usual partings by which it is divided into three benches. The least desirable coal, in quality, appeared to be the upper foot at the top of the seam. The lower two feet of the upper bench appeared well. The middle bench has a fine thickness of 5 feet 6 inches. There is a tendency, often seen elsewhere, in this seam to a more laminated structure of coal both at the top and bottom of this bench. These laminated layers will perhaps measure 15 inches for the upper and 12 inches for the lower, the latter having a slightly slaty tendency, resembling a similar horizon in the A. Sanders opening. As a whole, the coal is very fine and can not fail to be very popular and merchantable. No analyses have been made of the coal from this bank, but I think it safe to predict that it will serve an admirable purpose for the blast furnace. From this point, as well as from the entry on the A. Sanders land, a very large body of coal situated in the hills to the north and north-west, probably from 1,000 to 1,500 acres, is easily accessible. The general dip will be south-east, so that there

will be easy drainage of mines and an easy delivery of the coal to
the mouths of the mines.

The valley of the West fork in this neighborhood is sufficiently
wide to furnish a pleasant site for a mining village.

On the Middle fork of Sunday creek, we find, by a boring made
on the Abraham Post farm, north-east quarter, section 27, Monroe
township, that the coal maintains nearly its maximum thickness.
This is seen from the following section :

(A. POST'S BORING.)

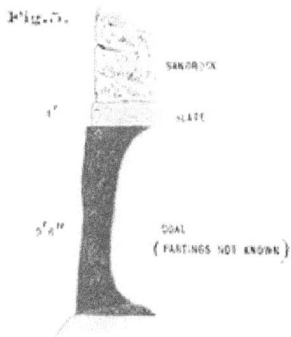

Fig. 5.

SANDROCK

1″   SLATE

5′ 6″   COAL
(PARTINGS NOT KNOWN)

The exact thickness of the partings could not be accurately
determined, but there is no reason to suppose that they are any
thicker than usual. We have a right also to infer the usual good
quality of the coal. We may also reasonably infer that the coal
extends in full thickness through the area lying in the angle
between the West fork and Post's. Indeed, this area of thick coal
may be believed to extend through all the south-western portion of
Monroe township, for we have found the coal along West fork on
the north ; at Post's, section 27, on the south-east ; and we know
that the seam is found in good development on Snow fork
of Monday creek, down the whole east line of Ward township,
Hocking county. There may be local disturbances and limita-
tions, but it is reasonable to suppose that the seam is continuous

throughout the area, on three sides of which we know it to exist. There has, quite recently, been a trial boring on section 7, Trimble township, about seven miles south of the Post boring, and on the line of the Atlantic and Lake Erie Railroad, which revealed 8 feet 4 inches of coal, at a depth of 80 feet below the level of Sunday creek. This very important fact warrants the presumption that the great seam extends in fine working thickness under a large part of Trimble township. This greatly enlarges the area of the productive coal-field. The following analysis of the coal from this boring was made by Prof. Wormley :

Specific gravity,	1.303
Moisture,	4.10
Ash,	5.50
Volatile combustible matter,	32.90
Fixed carbon,.	57.50
	100.00
Sulphur,	0.79
Sulphur left in coke,	0.49
Gas per lb. in cubic feet,	3.56
Ash, (color)	Dull white
Coke,	Compact

This shows a most excellent coal, and one well adapted to all the higher uses. It has good heating power, has little sulphur, and is rich in gas.

Having thus given facts and inferences relative to the extent of the great seam on the West branch of West fork and south of it, in Monroe township, I shall now trace the same seam farther up the branches of the West fork of Sunday creek.

On Rechter's branch of West fork, south-east quarter, section 18, Monroe township, the seam is found in magnificent development. It lies in the bed of the stream, and at no point has the stream eroded its bed deep enough to exhibit the lower bench of

the seam, but an excavation has been made by which I obtained
the measurements given in the following section :

(RECHTER'S, OR COAL, FORK,)

Fig. 6.

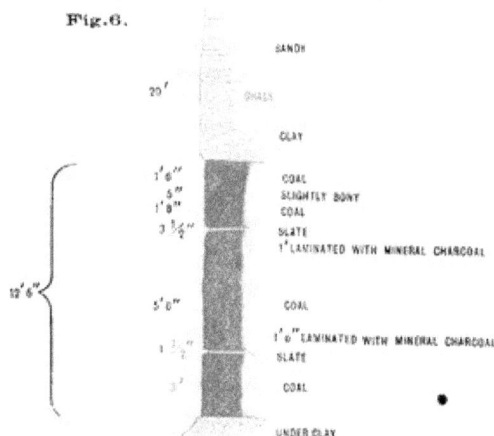

Here the total thickness of seam, from roof to floor, is 12 feet
6 inches. The two slate partings are respectively 3½ inches, and
1½ inch, leaving 12 feet 1 inch of coal.

In the upper bench there are 5 inches of coal slightly bony in
character, but from analyses, it is not found to be objectionable
for use, except in the increase of ash, which is 8 per cent. Three
analyses were made of four samples taken from the upper bench
to represent the whole of it : the average is as follows :

Specific gravity.................................... 1.295

Water,.. 4.76
Ash,... 6.50
Volatile combustible matter,.................... 32.23
Fixed carbon,.................................... 56.50

                                                   99.99
Sulphur,.. 0.91
Sulphur remaining in coke,...................... 0.29
Fixed gas per lb. in cubic feet................. 3.46

This shows an excellent quality of coal. It is not so rich in fixed carbon as other parts of the seam, but it will give satisfaction for all ordinary uses, and will also, I have no doubt, serve a good purpose for the blast furnace. It compares not unfavorably with the coal of the upper bench both in the Maginnis bank at Old Straitsville, and in the Straitsville Mining Company's bank at New Straitsville. At these points, the upper bench is very thick, and nearly two-thirds of the whole seam. The average of three analyses of the former, and two from the latter, as reported in the Ohio Geological Reports, are as follows:

No. 1, Maginnis bank, Old Straitsville, upper bench.
No. 2, Straitsville Mining Company's bank, New Straitsville, upper bench.
No. 3, Upper bench, Rechter's bank.

	No. 1.	No. 2.	No. 3.
Specific gravity,	1.267	1.269	1.295
Water,	6.32	6.25	4.76
Ash,	6.46	6.55	6.50
Volatile combustible matter	30.76	30.55	32.23
Fixed Carbon	56.46	56.65	56.50
	100.00	100.00	99.99
Sulphur,	0.80	0.87	0.91
Sulphur remaining in coke,	not giv'n	0.17	0.29
Permanent gas per lb., in cubic feet,		3.045	3.46

There is in the Rechter coal, less loss from water and 0.415 per cent. of a cubic foot more gas, the latter being a consideration of some consequence, since the New Straitsville coal is now largely used for gas-making. There would be also a small saving of expense in the purification of the gas since a little less of the sulphur enters it.

The middle bench of the Rechter fork coal, is 5 feet 6 inches in thickness. This part of the seam will, probably, from its high percentage of fixed carbon, be best adapted to the blast furnace, and yet, it may be found desirable to use both upper and middle benches together.

The average of five analyses is as follows :

Specific gravity	1.321
Water	5.16
Ash,	6.66
Volatile combustible matter,	28.84
Fixed carbon,	59.34
Total,	100.00
Sulphur,	0.81
Sulphur in coke	0.43
Permanent gas per lb., in cubic feet	3.34

The ash in this average, is rendered larger by including one of the analyses which gave an exceptionally large ash, and a quantity which does not, I think, prevail generally in this part of the seam. The fixed carbon, is large and determines the coal to have a high heating power. The loss in water is 5.16 per cent., which is considerably less than in the Straitsville coals. The sulphur as compared with New Straitsville (the whole seam, for it is all mined and used together,) is as 0.81 to 0.797 per cent. Of this sulphur 0.43 per cent. remains in the coke. This is an advantage if the coal is used for gas, as it reduces the expense of purification, but for the blast furnace, the less sulphur in the coke the better. But the amount is, after all so small, as not to create the slightest apprehension in the mind of the iron-master. The coke of the Youghiogheny coal retains, by Prof. Wormley's analyses, 0.66 per cent. of sulphur, and the coke of the widely used Connellsville, Pennsylvania, coal contains, even a larger percentage. Prof. Wormley finds in this coke, 2.17 per cent. of sulphur.

The coke of the New Castle or North Durham coal of England, more extensively used in the blast furnace than any other coke in the world, contains from 0.60 to 1 per cent. of sulphur.

On the land lately owned by George Welsh, southwest quarter, section 8, Monroe township, we find the seam of coal even thicker than on Rechter's fork. The following section shows the character of the seam:

(WELSH BANK,)

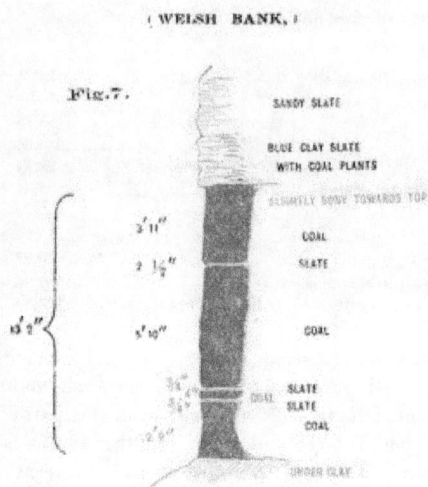

Fig. 7.

SANDY SLATE

BLUE CLAY SLATE
WITH COAL PLANTS

SLIGHTLY BONY TOWARDS TOP

3' 11"   COAL

2 1/2"   SLATE

5' 10"   COAL

3/4"   COAL   SLATE
3/4"   SLATE
                    COAL

13' 2"

.2' 9"

UNDER CLAY

Here the seam measures from roof to floor, 13 feet 2 inches, and from this only 4 inches are to be deducted for slate.

· The coal of the top of the seam, at this location has formerly been reached by stripping in the banks of the stream, but the inroads of water have always prevented the exposure of the lower portion. A small shaft sunk a little way from the stream, at the foot of the hill, has enabled me to examine the seam in minuteness of detail.

There are seen in the shaft 6 feet of blue slate above the coal. This slate is sandy at the top and more clayey at the bottom. Some beautiful impressions of coal plants were seen. The slate will furnish a strong and impervious roof. The upper bench, nearly 4 feet thick, (exactly 3 feet 11 inches), shows entire freedom from visible sulphur in the form of bisulphide of iron, (pyrite), and probably what little sulphur there may be exists chiefly as an organic compound. The tendency which the upper bench often shows, both at Straitsville and on Sunday creek, to a cannel-like, or, as some miners would term it, a bony structure is seen here. This kind of coal is in thin streaks and will, I think, do little injury to the coal, should the seam be mined as a whole. It may increase somewhat the percentage of ash, but, if the coal is used for gas, I should expect it to increase the illuminating power of the gas. For household use the coal will, I think, prove to be a brilliant fuel for the parlor grate.

The middle bench of 5 feet 10 inches is remarkably uniform in character, (excepting the bottom part which is somewhat more earthy,) and exceedingly fine in quality, and the same may be said of the lower bench, so far as quality may be determined by the analysis, and by a careful examination of the pile of coal taken from the shaft. I could nowhere, either in the shaft or in the coal thrown out, detect with the eye the slightest appearance of bi-sulphide of iron. The coal breaks out in large firm blocks and will evidently bear transportation and handling with very little breakage and waste. Such coal can not fail to command the highest price in any market.

It is a little remarkable that at this point, where, so far as I know, we have the *maximum development of any coal seam in the State of Ohio*, we have also, *coal of the very finest quality.*

The following analyses have been made of five samples of coal from the Welsh bank, representing the different portions of the seam :

No. 1, middle of the upper bench.
No. 2, upper part of middle bench.
No. 3, middle of middle bench.
No. 4, lower part of middle bench.
No. 5, middle of bottom bench.

2

	No. 1.	No. 2.	No. 3.	No. 4.	No. 5.
Specific gravity,........	1.302	1.316	1.300	1.385	1.312
Moisture,.................	4.60	5.20	4.30	4.90	4.40
Ash, ...	4.70	5.00	4.20	13.30	2.70
Vol. combus. matter,....	33.40	31.40	32.70	28.30	30.60
Fixed carbon,............	57.30	58.40	58.80	53.50	62.30
	100.00	100.00	100.00	100.00	100.00
Sulphur,.................	0.71	0.74	0.71	0.79	0.90
Sulphur left in coke, ...	0.35	0.38	0.35		0.43
Gas per lb. in cub. feet,..	3.72	3.48	3.48	3.24	3.72
Ash, ...	Yellow.	Yellow.	Yellow.	Dull w'te	Dull w'te
Coke,....................	Compact	Com.	Com.	Pulv.	Com.

These analyses prove the very superior quality of the coal. No. 4 gives an exceptionally large ash, but the sample came from the bottom of the middle bench, and should coal of this character prove to have any considerable horizontal extent through the mine, which I very much doubt, it might easily be excluded in mining. If thus excluded, there would be left more coal than is found at any other mine in the State beyond the limits of Sunday creek valley.

Higher up the stream, on Rogers' fork of Sunday creek, near the north line of the north-west quarter, section 7, Monroe township, the great seam has been reached by boring.

The coal was found in fine thickness, the measurement showing 12 feet. The partings of slate, which from the borings are doubt- less as thin here as elsewhere, were not exactly determined in position. The borings of coal brought up in the sand pump are bright and handsome, and, so far as I could judge from pieces so small, the quality of the coal is not inferior to that found at other points. The seam is about 28 feet below the level of the stream. The relative thickness of this seam may be seen from the following section :

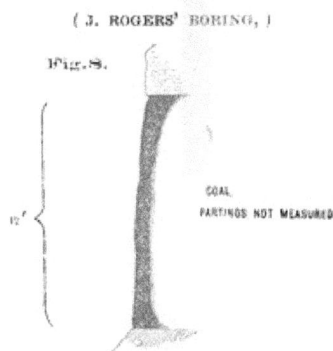

(J. ROGERS' BORING,)

Fig. 8.

COAL
PARTINGS NOT MEASURED

12'

On the Middle fork of Sunday creek the great seam is found in full development. Reference has already been made to a boring on the Abram Post farm, on the north-east quarter of section 27, Monroe township, where the coal was found to be 9 feet 6 inches thick. At this point the seam is about 60 feet below the surface. There is a foot of slate directly over the coal, and a heavy sand- rock 41½ feet thick above the slate.

On Dodson's run, a branch of the Middle fork of Sunday creek, a boring was made on the William Fisher place, in the north-west quarter of section 23, Monroe township, and the great seam was found to be 10 feet 10 inches thick. The distance below the surface is 53 feet. Directly over the coal are 4 feet of slate, and over the slate 37 feet of sandrock.

(FISHER'S BORING,)

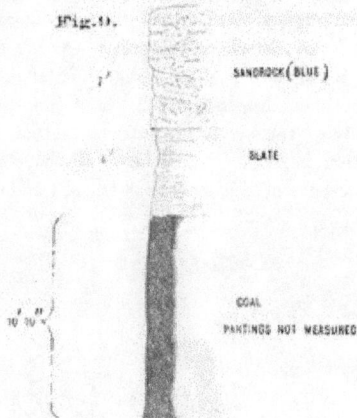

Fig. 9.

SANDROCK (BLUE)

SLATE

COAL
PARTINGS NOT MEASURED

I did not see any of the borings from this point, but I have no reason to doubt the good quality of the coal.

On the Nelson Rogers farm, at Ferrara, in south-east quarter of north-west quarter of section 15, the great seam was struck by boring at a reported depth of 29 feet below the surface of the bank near the bridge by the Rogers house. The seam was found to be 11 feet 6 inches thick.

(NELSON ROGERS' BORING,)

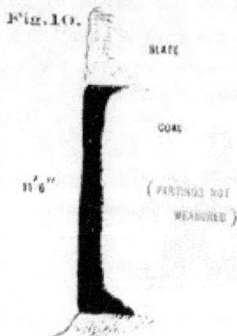

Fig. 10.

SLATE

COAL

11' 6"   (PARTINGS NOT MEASURED)

A mile or more up the stream (Middle fork) we find the Sands' bank, in the north-east quarter of south-east quarter of section 9, Monroe township.   Here the stream has exposed the upper portion of the seam, and the coal has been mined in a limited way for many years.   An entry has been made for 30 or 40 yards, and the mine drained by a steam pump.

The following is a detailed section of the

SANDS' BANK,

Fig. 11.

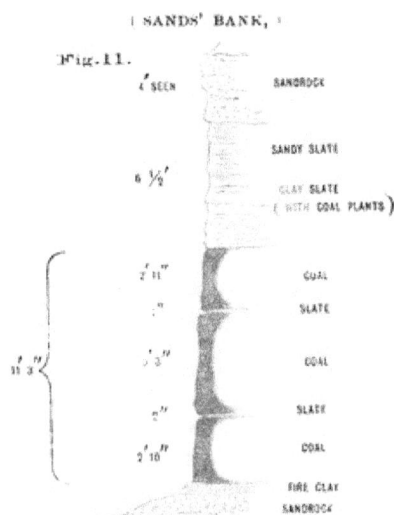

4' SEEN	SANDROCK
6 ½'	SANDY SLATE
	CLAY SLATE (WITH COAL PLANTS)
2' 11"	COAL
"	SLATE
3"	COAL
"	SLATE
2' 10"	COAL
11' 3"	FIRE CLAY
	SANDROCK

The coal is of excellent quality, and has been long prized by blacksmiths who come fifteen or twenty miles in wagons over hilly roads to obtain it.   There is a foot of the bottom of the middle bench which appears to be more slaty than any other portion, but for all ordinary uses it will not probably be rejected.

Analyses of seven samples representing the seam from roof to floor were made by Prof. Wormley, and published in the Ohio

Geological Report for 1870. The average of these seven analyses
is as follows:

Specific gravity,..	1.300
Moisture, ...	6.42
Ash,...	5.54
Volatile combustible matter.................. ...	33.87
Fixed carbon,.....................................	54.17
Total, ...	100.00
Sulphur,..	0.88

The percentage of sulphur of the upper 7 feet of coal is only
0.53, which is quite small.

The coal, as a whole, is of excellent quality, and is adapted to
all the higher uses both of gas and iron making. While at this
exposure the average percentage of fixed carbon is not large, yet,
I have no doubt that this is exceptional, and I should confidently
expect that if another opening were made upon this very valuable
estate, or if the present entry were driven farther, the coal would
in all respects authenticate itself as equal to any in the valley.

I have thus given a pretty minute description of the coals of the
great seam so far as I could find exposures, and so far as the
chemical analyses extend. In regard to the latter, while they
are more complete than have ever been made of any limited coal
area in the country, so far as I know, yet, for the fullest scientific
discussion, I could have desired a little more special analytical
investigation of the coal with reference to special uses.

This Sunday creek valley coal-field, as a whole, shows a coal
of remarkable purity and excellence. It must be ever remembered
that there are in all coal fields considerable variations of quality,
and while here, at some points, the coal may be less good than at
others, this distinction is only applicable to very limited areas. It
is often the case in mines, that the coal in a single entry may be
quite inferior to the rest of the coal. Hence, if at some point in

the Sunday creek valley whence samples have been taken there may be a little more sulphur, or less fixed carbon, or less permanent gas, it must be borne in mind that these cases, not conforming to the general average, are only of local and exceptional character. Should a company owning 1,000 or 2,000 acres of the land containing the thick coal, chance to have an exposure where the coal is not apparently the best in this field, this circumstance is not to be allowed to throw the slightest suspicion upon the general average quality on its estate, since its value is determined by the character of the whole Sunday creek territory, which includes the several company estates.

So important do I regard this view of the matter, it being one I have often seen verified elsewhere, that I have prepared a table showing the analyses from the different openings on Sunday creek, and the general average of the whole, to which I append, for comparison, analyses of several coals from other locations, chiefly such, however, as are used for iron-making. This report of the Upper Sunday creek valley covers a very large area, and I have given not selected facts or preferred analyses, but *all the facts* and *all the analyses*. I have given, also, the average of *all the analyses for the whole field*. This is a very severe test, for all intelligent men know that in so large an area, there will be exceptional localities where the coal is of less desirable quality. While on Monday creek, the coal at New Straitsville shows in its analysis a very fine quality, and at Old Straitsville an excellent quality also, though less rich in fixed carbon, yet, the analyses of the coal of the same thick seam a mile or less south of New Straitsville, shows a marked inferiority, with an average of 53.28 per cent. of fixed carbon, and 1.79 per cent. of sulphur.

If this result were grouped with the averages at Old and New Straitsville, as parts of a common field, it will be readily seen what effect it will have upon the general average. But having served the Upper Sunday creek field, now owned by the several companies for whom this report is prepared, in this way, by including the worst with the best, and giving the average of 27 analyses,—all ever made,—it is certainly just to quote from the State Geological Report the average of the 15 analyses of coals taken from a less area in the Monday creek valley, in the neighborhood of Straitsville. I am inclined to think that if in the Briar Hill coal district of Mahoning county, a similar method were pursued of giving all the variations of quality in an area as large as that of the Sunday creek field, the average of 27 analyses representing all parts of the seam, would present a different character from that of my table where we have the analyses of only three samples of this standard coal, and these possibly selected with some care. The samples of Sunday creek coal analyzed, were all selected by myself, and my sole aim was to secure such samples as most fairly represented the seam. In some cases each foot of coal in the vertical range is represented by a sample.

The Jackson shaft coal, No. 15 of the table, has authenticated itself as an excellent coal to be used, in the raw state, in the blast furnace. The same may be said of the Ashland coal, No. 16 of the table. Of the Brazil block coal, from Indiana, we have only a single analysis by Prof. Wormley. This gives a relatively small percentage of fixed carbon and I have appended in a note the average fixed carbon of several analyses found in the Indiana Geological Report for Clay county, Indiana. As none of these atter analyses give any sulphur whatever,—an unaccountable omission—I have not thought it best to insert the whole of these analyses.

The following tables are worthy of careful examination, and show the relative character of the Sunday creek coal much better than could be done by any verbal statement :

	SUNDAY CREEK.						MONDAY CREEK.			
	No. 1.	No. 2.	No. 3.	No. 4.	No. 5.	No. 6.	No. 7.	No. 8.	No. 9.	No. 10.
Specific gravity,	1.290	1.295	1.303	1.325	1.290	1.300	1.289	1.300	1.278	1.291
Moisture,	5.02	1.95	5.66	4.09	6.45	5.24	6.05	6.84	5.60	6.50
Ash,	5.03	5.25	6.78	5.96	5.34	5.05	4.65	4.71	4.97	4.78
Volatile combustible matter,	29.82	37.06	30.96	31.28	33.42	33.46	34.25	31.55	36.90	33.78
Fixed carbon,	62.40	59.75	57.98	58.99	54.77	56.17	55.05	56.02	52.50	54.91
Total,	100.00	100.00	99.96	99.38	100.00	100.00	100.00	99.56	100.14	100.03
Sulphur,	0.75	0.85	1.20	4.77	0.49	0.00	0.75	0.52	1.25	1.16
Sulphur remaining in coke,		0.54	0.54	0.57			0.11			
Permanent gas,		3.00	3.00	3.95		3.45	3.18			

No. 1. Coal from Bessemer Workers' bank.
No. 2. Average, 3 samples, 3 benches' bank.
No. 3. Average, 3 samples, Kachler's bank.
No. 4. Average, 3 samples, Straitsville bank.
No. 5. Average, 3 samples, Randel bank.
No. 6. Average of No. 4 above.

No. 7. Average, 4 samples, New Straitsville.
No. 8. Average, 3 samples, Maginnis' old bank, old Straitsville.
No. 9. Average, 3 samples, Deer Lick mine.
No. 10. Average of Nos. 7, 8, and 9.

	BRIAR HILL, O.							
	No. 11	No. 12	No. 13	No. 14	No. 15	No. 16	No. 17	No. 18
Specific gravity	1.284	1.298	1.059	1.289	1.067	1.304	1.178	1.208
Moisture	3.68	2.47	3.90	2.93	7.59	6.65	3.96	6.54
Ash	1.16	1.45	6.66	5.67	9.10	4.51	1.96	3.49
Volatile combustible matter	32.24	31.83	29.10	31.17	50.90	34.54	35.74	34.73
Fixed carbon	62.96	64.25	60.49	62.44	32.80	54.28	58.80	54.29
Total	100.00	100.00	100.00	100.00	100.00	100.00	100.00	100.00
Sulphur	0.85	0.06	0.82	0.74	0.74	1.07	0.29	0.89
Sulphur remaining in coke		0.48	0.04		0.72			
Permanent gas			3.52					

No. 11, Briar Bell, Cleveland Ridge, 1 sample.
No. 12, Youngstown, 1 mine's sales, 1 sample.
No. 13, Mahoning Co., Wideman's mine, 1 sample.
No. 14, Average of Nos. 11, 12, and 13.

No. 15, New Fire Shaft coal, Jackson, O., 1 sample.
No. 16, Ledges of Ashland, Ky., Pot, coal, 1 sample.
No. 17, Sample of Brush Carbon, coal, each sample multiplied by Prof. Wormley.
No. 18, Average of Nos. 15, 16, and 17.

† By adding 1 analyses to late White, the average fixed carbon is 55.08.
§ Partial analyses of these made from Clay Co., Indiana, reported in Indiana Geological Report part 27-28.

It will be seen that, in the tables, I have given the analyses of nearly all the bituminous coals of the West, used in the raw state in blast furnaces. The Briar Hill coal of the Mahoning region is a coal of great excellence, as the table shows. But this seam is relatively thin, and the coal is mined at considerable expense. It is believed that your advantages of thickness of seam and cheapness in mining, will enable you to compete successfully in markets not very far from the Mahoning coal fields. The best of the Briar Hill coal will all be needed for iron-making, and should be husbanded for that special use, and this is the view taken by the most intelligent men of that part of the State. Hence, when the best facilites of transportation to Cleveland and other important points in that direction are secured to you, as they soon will be, I can see no reason why a very large trade may not spring up in that region heretofore supplied from other sources.

The other coals given in the table, viz.: the Jackson shaft coal, Ashland, (Kentucky) coal and Brazil, (Indiana) coal, are coals of great excellence and value, and have so authenticated themselves when intelligently used in the blast furnace, but none of them have the same wide range of uses with the Perry county coals. They will, however, all be needed in the development of vast iron manufactures in which this country is certainly destined to become the foremost nation of the world. The locations of these coals are quite too remote from the Sunday creek field to admit of much rivalry of competition. With proper railroad facilities all our coals of the better class in the West will find remunerating markets.

I have thus given in minuteness of detail, the structure of the great Sunday creek seam of coal, at such points as I found it accessible, and upon reports of borings which I regard as truthful. The seam in its maximum development, extends through many thousands of acres, and will afford a supply of coal to meet present and prospective markets for several generations to come.

The seam, thinned to less proportions, is found to have a wider extension and I propose to trace it beyond the limits of the Sunday creek valley. As north of Straitsville and Shawnee it grows thinner, so north of the Sunday creek valley it is found to extend to New Lexington in diminished development, but generally maintaining a good working thickness. As a general rule, the quality is less good as we trace the seam to the north, but there may be exceptions to this.

In the valley of the Moxahala, the chief Sunday creek seam is found to be persistent and generally of sufficient thickness for profitable working, and of fair quality. Here it is popularly called the Upper New Lexington seam of coal.

At the bank of James Fowler, in north-west quarter section 30, Pleasant township, I find the seam measuring 5 feet. There are two very slight partings, one 1 foot 10 inches from the bottom, and the other 3 feet 1 inch. The coal appears well, and resembles the Sunday creek coal in its physical structure. Over the coal were seen from 5 to 6 feet of clay slate. On the Thomas Kinsell land in south-west quarter, section 35, Pike township, I found the seam to measure 5 feet $1\frac{1}{2}$ inches, from which are to be deducted two slate partings measuring 2 inches and $\frac{1}{2}$ inch respectively. The coal appears well. There is a blue slate roof above the Kinsell bank. The following is a section.

(KINSELL'S BANK,)

Fig. 12.

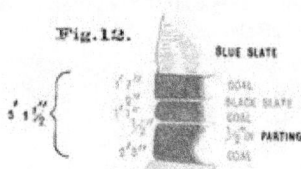

BLUE SLATE
COAL
BLACK SLATE
COAL
$\frac{1}{2}$ IN PARTING
COAL

The same seam is found, having about the same thickness, on the lands of James Sheeran, south-east quarter, section 23, Pike township, William Wiggins, south-west quarter, section 24, Pike township, and Abram Parkes, section 22, Pike township. The measurement on the latter farm, near the railroad on the south fork of the Moxahala, is a little less thick, the section being as follows :

	Feet.	Inches.
Clay slate,................(seen)...................	3	0
Coal, upper bench,.......................................	1	3
Slate parting..	0	3
Coal, middle bench,................	0	9
Slate parting,..	0	1
Coal, lower bench,...................................	2	1

There is apparently **more bi-sulphide of iron (pyrite) disseminated** in fine streaks through the coal **here than at any other exposure examined** in the Moxahala Valley.

The seam extends through the dividing ridge between the waters of Moxahala and Rush creeks, and is seen on the Gordon farm, **in** the cut **of the approach** to the tunnel. A general section of the rocks at **this point is** given, as it shows not only the position of the coal, but **of the heavy** sandstone rock in which the tunnel is excavated.

Iron Ore, "Sour Apple" Seam—	Feet.	Inches.
Not exposed,..	7	0
Heavy sand rock (the floor of the tunnel at the north end being about 10 feet above the bottom),............................	45	0 by estimate.
Slate, sandy at top, clayey **at bottom**......	12 to 15	0
Coal, upper bench,................................	1	4
Slate,...	0	3
Coal, middle bench,...............................	1	0
Slate,..	0	1
Coal, lower bench,................................	2	5
Under clay,...	4	0

The upper bench is highly laminated with mineral charcoal. This seam, throwing out a few inches at the bottom of **the lower bench, found to** be impregnated with bi-sulphide of iron, has been used somewhat for iron-making in the Zanesville furnace. For other uses, its general excellence and dry burning **quality will** make the seam, as a whole, acceptable.

There is in **all the** hills bordering the Moxahala a vast body of of coal. It **must be remembered** that **the** majority **of the seams** of bituminous coal are less thick than this seam, as developed on the waters of Moxahala. Four and a half to five feet constitute an easy and convenient working thickness, and, generally, a seam of about this thickness is preferred by miners.

OTHER SEAMS OF COAL.

Having given a detailed account of the range, thickness and character of the great Sunday creek seam of coal, the way is prepared to notice other seams upon these lands. There are three besides, one below and two above the great seam. The one below is generally called the lower New Lexington seam. East and north-east of New Lexington, this seam is of considerable economic value, and has been considerably mined. The general quality of the coal is good. Nowhere in the valley of Sunday creek have the streams cut their beds sufficiently low to expose this coal. Its place ranges from 20 to 25 feet below the great seam. This interval, however, varies somewhat. In the Moxahala valley, this seam gives proof of its existence in many places, and it may prove when opened, to be sufficiently thick for profitable working. The seam is often very uncertain, and one is not quite sure of it until he has actually proved its existence at a given point. Probably little money will be expended upon the working of this seam upon your lands while you have such vast quantities of other and better coal.

THE MIDDLE OR NORRIS SEAM.

This seam is found in place from 40 to 50 feet above the great seam. The best development of it, so far as shown by exposures, is at the old Norris bank near Millertown in section 21, Monroe township. Here it measures 6 feet in thickness.

The following section shows the seam :

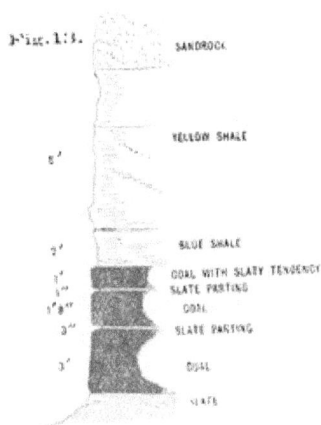

Fig. 13.

SANDROCK

YELLOW SHALE

BLUE SHALE

COAL WITH SLATY TENDENCY
SLATE PARTING
COAL
SLATE PARTING

COAL

SLATE

The quality of the coal is fair, but greatly inferior to that of the great seam. There is considerable sulphur (as bi-sulphide of iron) visible in the upper 10 inches of the lower bench, and also a tendency to slate in the top of the seam. It is not so dry-burning a coal as that of the great seam, but it is not highly bituminous and caking. It does well in the grate and will be a fair steam coal.

On the Grigsby farm, section 9, Monroe township, the middle seam has been somewhat opened and mined for neighborhood use. Here it measures 4 feet in thickness. Two samples of the coal from this bank were analyzed by Prof. Wormley and the result published in the first Annual Geological Report (p. 119), as follows :

	No. 1.	No. 2.
Specific gravity,	1.277	1.350
Water,	3.80	3.80
Ash,	4.60	6.30
Volatile matter, combustible,	38.80	37.00
Fixed carbon,	52.80	52.90
	100.00	100.00
Sulphur,	3.59	4.89
Cubic feet gas per lb.,	3.03	3.08

The sulphur is pretty large, in percentage, but in other respects the coal is fair and could be used where the sulphur would not be objectionable. The time will come in the future development of our industries when this coal will serve an important purpose, as similar coals now do in Great Britain, in being used with a wise economy for less specialized uses, and thus permit the saving of the best coals for furnaces, gas works, &c., &c. The best coals of the old world are very carefully husbanded. The middle or Norris seam is found on the Neesly McDonald farm, section 22, Monro township; at J. B. Latta's, section 4, Pleasant township; at J. Pyle's, in Pleasant township; at Benjamin Sanders', Monroe township; above the great seam at the Sands' bank, section 9, Monroe township; at Moxahala village, and at many other points.

At the "Sands bank" the middle coal has been well opened, and measures 4 feet 2 inches, with 4 feet of clay slate roof. No slate partings were seen, and the coal appeared to promise well. Here the interval between this seam and the great one below is 50 feet.

At Benjamin Sanders', on West fork, the middle seam measures only 2 feet 6 inches, and the quality is poor.

At Ferrara, on the Nelson Rogers farm, the place of the middle coal is seen, but only a few inches of coal are found.

The middle seam appears in the hills near Moxahala village, and at one exposure measured 4 feet 2 inches, with a 2 inch slate parting a little above the middle.

On Thomas Kinsell's farm it is thinner, measuring only 2 feet, with 2 inches of slate in the middle. The upper bench is quite sulphurous. This is 40 to 45 feet above the place of the great seam. In the hills northward, toward New Lexington, the middle seam is not often found, it being replaced by the heavy sandrock almost everywhere found over the "great seam," here called the Upper New Lexington seam.

UPPER OR STALLSMITH SEAM.

This seam of coal is found in most of the hills bordering the Upper Sunday creek valley, its stratigraphical place being, by the measurements, from 25 to 45 feet above the middle seam. It is probable that the larger measurements are inaccurate, sometimes from the disturbance of the barometer, and sometimes from confounding the Stallsmith seam with another which is occasionally found a few feet above it. There is a good development of this seam on the Benjamin Sanders farm on West fork, where it has been mined and used by Mr. Sanders, for domestic purposes in preference to the dryer coal of the "great seam," which is well developed on his land. Mr. Stallsmith, on the hill south of Mr. Sanders', has mined this seam for his own and neighborhood use during many years. The seam everywhere affords a hard flinty coal, highly resinous and caking, and it is because it is so "fat and oily" in burning that many prefer it.

The following is a section of the coal as exposed on Mr. Sanders' place :

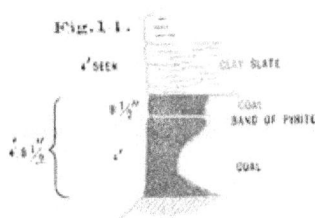

Fig. 1 1.

4' SEEN — CLAY SLATE

8 1/2" — COAL
BAND OF PYRITE

4' — COAL

4 8 1/2"

3

At the opening there is seen a thin band of bi-sulphide of iron 8½ inches below the top. This may not extend through the hill. The purest coal is apparently that above the sulphur band. The 4 feet below are hard, bright, compact coal, and no parting of any kind was observed in it. Careful inspection will reveal here and there traces of bi-sulphide of iron which will impair the value of the coal for the higher uses. Some of this coal has been coked by Mr. Thomas N. Black, on the ground near the opening, and the coke is hard and firm. Nothing but the sulphur can prevent this coal from being a very important addition to the mineral wealth of the Sunday creek valley, and it is possible that at other locations there will be found less of this undesirable constituent.

The following analyses have been made of this coal by Prof. Wormley :

No. 1, Sanders' bank.
No. 2, Stallsmith's bank.

	No. 1.	No. 2.
Specific gravity,...	1.294	1.254
Moisture,.. ...	DRIED AT 212°	3.80
Ash,..	2.80	4.14
Volatile combustible matter,............................	41.70	40.21
Fixed carbon,...	55.50	51.85
Total, ...	100.00	100.00
Sulphur, ...	2.56	2.62
Permanent gas per lb. in cubic feet,......................		

The Stallsmith seam is opened on the land of E. Springer, south-east quarter, section 13, Salt Lick township, also on the land of E. Alderman, south-west quarter, section 19, Monroe township. On the land of Morgan Devore, south-east quarter, section 10, Monroe township, the seam measures 3 feet 6 inches, and shows no slate parting. On the hill above the Sands bank, (of the "great seam"), the Stallsmith seam has been opened and measures 3 feet 6 inches. The same seam is seen on the land of M. Longstreth, north-west quarter, section 4, Pleasant township, where it measures 4 feet. On the land of B. Green, south-east quarter, section 33, Pleasant township, the same seam is reported to measure 4 feet. The same seam is credibly reported to measure 5 feet on the Grannan farm, north-west quarter, section 33, in the same township. The blossom of it is seen at many other points, but the seam has not been opened. As the quality of the coal of a seam often shows very great diversity, it is possible, and quite probable, that this Stallsmith coal will be found, at some location on these very extensive properties, to possess such a degree of purity as will make it a very desirable coking coal. A good and cheap coke is a great desideratum in central Ohio, most of the coke now used being brought from Pennsylvania. The Stallsmith coal, being very rich in gas and caking in character, will be preferred for household use by a large number whose ideas of excellence in coal have been derived from the use of the more caking coals of western Pennsylvania. As a steam coal it can not fail to serve an excellent purpose.

There are above the Stallsmith seam two other seams in the Sunday creek hills, but they are nowhere opened, and probably are not of sufficient thickness for profitable working.

One of these is seen on Joseph Rogers' hill, about 27 feet about the Stallsmith seam. This, however, may be a slip, and the true place may be above. On the Benjamin Green and Street farms a very thin seam is seen, about 45 feet above the Stallsmith seam. Another very thin seam is often found about 100 feet higher. There is probably an intermediate seam which may be entirely local. These upper seams are so thin and valueless, that, with a single exception of the upper one of all, no excavations were made to determine their character. In the exceptional case the coal was not only thin but worthless.

IRON ORE.

Considerable quantities of iron ore are found in the Sunday creek valley. The lowest ore, in the stratigraphical series, is found a few feet (generally 8 or 10) below the great coal seam. But as the great seam lies very low in the Sunday creek valley, the only exposure of this ore is in the bed of West fork, where the bed of the stream is below the coal. The ore is a pretty rich siderite, (or blue carbonate of iron,) sometimes slightly oxidized on the outside. This ore is in nodules, and does not constitute a solid layer, and it is doubtful whether it could be mined by drifting, (the only practicable method on your property,) without too great expense. On Snow fork of Monday creek, in Ward township, a similar ore is found in the same geological horizon, viz: about 9 feet below the great seam of coal. A sample of the ore from Snow fork, was analyzed by Prof. Wormley, and the result published in the first Annual Report of the Ohio Geological Survey, from which I copy as follows :

Iron ore 9 feet below the great seam of coal on Snow fork:

Specific gravity,	3.200
Protoxide of iron,	37.22
Sesquioxide of iron	3.64
Manganese,	1.20
Alumina,	0.60
Lime,	2.40
Magnesia,	2.16
Foreign matter,	18.82
Carbonic acid	28.10
Sulphuric acid,	trace
Phosphoric acid,	trace
Combined water,	5.70
Loss	2.56
Total,	100.00
Metallic iron,	31.50

This is a remarkably pure ore, containing only chemical traces of the two most deleterious elements in iron ore, viz : sulphur and phosphorus. From the appearance of the ore of the same geological horizon on West fork, I should think it would yield from 5 to 10 per cent. more metallic iron than that from Snow fork.

West of Straitsville on the tops of the high hills west of Monday creek, the ore of this horizon is found in a regular layer, and here it has been changed, through long exposure to atmospheric agencies, to a limonite or hydrated sesquioxide of iron. It has been considerably mined by stripping and taken by railroad to the Columbus furnace.

It is possible that this ore might be found in sufficient development for profitable working on the Moxahala, or in the hills farther north towards New Lexington. In the very few exposures I have found of the strata immediately below the coal, (there called the Upper Lexington seam,) I have nowhere observed the ore. In the vicinity of the Del Carbo mines the ore of this horizon is seen in nodules of considerable size.

Sometimes we find in the shales above the great seam on Sunday creek, large nodules of siderite. They are irregularly disseminated, and not promising in richness of iron. In the same shales at one or two points, especially on the Snyder farm, are two layers of nodules of richer siderite, respectively 16 and 21 feet above the great seam of coal. The lower is about 3 inches thick and the upper 4 inches.

Near Millerstown a layer of siderite 5 inches thick is found about 4 feet below the middle or Norris seam of coal.

"Sour apple ore." About 15 feet above the Norris coal is generally found a stratum of limonite ore, ranging from 6 to 13 inches in thickness. Near the outcrop it is thoroughly oxidized, and is a true hydrated sesquioxide of iron. I have traced this ore in its proper geological horizon from Sunday creek to New Lexington. In Pike township it is in larger development, and will afford a large quantity of valuable ore. On the farm of Wesley Moore, north-east quarter, section 14, Pike township, the ore has been taken out to a slight extent, enough however to show a very valuable deposit. By barometer this ore is 63 feet above the upper New Lexington or Sunday creek seam of coal. The ore is in large nodular masses imbedded in clay. The section here shows a range of these nodular masses 2 feet 8 inches thick, which if they constituted a compact layer would make a thickness of perhaps from 10 to 14 inches. At the outcrop the ore is pretty well oxidized, and is a limonite of fine quality, but farther in the hill and under more impervious strata it retains its original character of a blue carbonate or siderite. Directly over the ore at Wesley Moore's are 4 feet of white clay. The same seam of ore is reported to be dug to a considerable extent on the high ground north of Rush creek in a direction north-west of New Lexington. Vast quantities of this promising ore can be obtained in this region.

An analysis of the "sour apple ore" obtained on the Latta farm on Sunday creek was made by Prof. Wormley, and given in the first Annual Geological Report. I do not regard this ore as equal in quality to the ore of the same seam found in Pike township. The following is Prof. Wormley's analysis:

Specific gravity,	2.714
Combined water,	8.90
Silicious matter,	25.60
Sesquioxide of iron,	59.03
Alumina,	1.56
Phosphate of alumina,	0.59
Oxide of manganese,	2.40
Phosphate of lime,	1.10
Phosphate of magnesia,	0.70
Sulphur,	trace.
Total,	99.88
Metallic iron,	41.31
Phosphoric acid,	1.21

For certain purposes of mixture the phosphorus in this ore will be unobjectionable.

A better quality of this ore was seen on the Nelson Rogers farm.

THE LATTA ORE.

This is a blue carbonate of iron, or siderite found about 15 feet above the Stallsmith seam of coal. It is somewhat earthy, and has little tendency to change, on exposure in its outcrop, to the oxidized or limonite variety. It is often found in very large blocks, but at no point have I found positive proof that it forms a continuous stratum. On the Latta farm some of these blocks are 2 feet thick and weigh several tons. On the Kelita Rogers farm there are in one place three layers of large nodules of ore within a vertical distance of 4 feet 6 inches. These nodules measured, in their several layers, 6, 14 and 13 inches respectively.

The same ore is found on the farms of Benjamin Green, Harrison Grigsby, A. S. Alderman and John Pyle, and doubtless may be obtained at its proper horizon in all the hills in that region. The ore is not very rich in metallic iron, Prof. Wormley obtaining from the sample taken from the Latta farm 26.12 per cent., and from that from the Rogers farm 23.78 per cent. The ore of this percentage will only be available as a mixture with richer ores.

On the John Street farm, south-east quarter, section 28, Pleasant township, there is a layer of large nodules of blue carbonate of iron, often considerably oxidized on the outside. There are 2 feet of these nodules, probably enough to make a compact and continuous layer of 1 foot. The ore appears to be somewhat better than the Latta ore, and yet I am inclined to think it the geological equivalent of the Latta ore. Subsequent investigation might determine this point.

There is considerable ore in nodular form scattered through the shales above the Street ore already described. On the Benjamin Green farm, near the old Alton mill, there is seen by the roadside a layer of nodules of an earthy blue carbonate of iron, the nodules being from 8 to 10 inches thick. Fifteen or eighteen feet above this ore is a very thin stratum of coal, and 15 feet higher an earthy ferruginous fossiliferous limestone. This limestone is, probably, the equivalent of one

found 54 feet below the coarse ferruginous limestone near
the barn on the Devore hill. It is possible that the
highest fossiliferous limestone on the hill west of Street's is
the equivalent of the Devore limestone. If this is so, the
place of the Stallsmith seam of coal will be above the bed of
the stream in the valley by Mr. Street's. By the barometer
this valley is 25 or 30 feet lower than the more eastern valley
by Alton's mill. By comparison of the geological sections it
appears probable that the ore above the Alton mill is about
15 feet above the place of the larger or Latta seam.

Should the Sunday creek coal be employed for iron-making,
there is no propriety in limiting the resources of ores to those
I have already described. Furnaces will necessarily be
established at such points as will combine the larger number
of the needed materials within reasonable proximity. Furnaces established, for example, at points on the Moxahala or
Rush creek, where there are undoubted supplies of water,
would command all the Coal-measures ores found sufficiently
near all the railroads in Perry and Muskingum counties.

Superior ore is found at many points along the line of the
Cincinnati and Muskingum Valley Railroad, and also on the
Newark, Somerset and Straitsville Railroad. All of which
may be made available.

I am convinced that there is a vast quantity of excellent
ore in the lower portion of the Coal-measures to be furnished
as soon as there is a demand for it.

For example, in addition to the so called " sour apple " seam,
so well developed about New Lexington, we have, near
Wolf Station, ores from above the flint stratum, a considerably lower geological horizon. Farther west in the neighborhood of Bremen, on the adjacent higher grounds, we have
a very fine ore near the base of the Coal-measures This
latter ore has been analyzed by Prof. Wormley, and is found
to contain 51.66 per cent. of metallic iron, with only 0.19 per
cent. of phosphoric acid and a trace of sulphur. This, considering its richness and purity, is one of the best ores of the
State. This ore can be delivered cheaply to any furnace
located between Lancaster and Zanesville.

It is an interesting geological fact, that the Moxahala in its eastern course has eroded its valley below the Coal-measures; the stream, with its branches, near Newtonville, flowing upon the Newtonville limestone, the equivalent of the Maxville limestone, a formation which always reposes upon the Waverly. This brings the strata in all the surrounding hills into the range of the lower Coal-measures and these lower measures generally contain the more important seams of iron ore. On the high grounds north of the Moxahala there will doubtless be found much ore. On the National road, in Hopewell township, ore of superior quality is now mined and taken by wagons to the Zanesville furnace. The Riffle ore near Gratiot gives, by Prof. Wormley's analysis, 52.51 per cent. of metallic iron, with only 0.38 per cent of phosporic acid, and a trace of sulphur. From these and similar facts I infer that furnaces on the Moxahala might obtain a vast quantity of excellent ore.

This discussion of the coal deposits of the Upper Sunday creek and Moxahala valleys, would be incomplete without some consideration of the facilities for shipping coal to the great markets of the North and West. The distances to the more important points of consumption and distribution are given in the following table:

TABLE OF DISTANCES.

Columbus to Ferrara,..59   miles.
Columbus to Straitsville,.. 63   "
Columbus to Nelsonville,... 62   "
Toledo to Ferrara, via A. & L. E. R. R...........................174   "
Toledo to Straitsville, via Columbus (East Line.)................183   "
Toledo to Straitsville, via Columbus (West Line.)...............190¼   "
Toledo to Shawnee, via Monroeville and Newark,..................198   "
Cleveland to Ferrara, via Oxford and Dresden,...................164   "
Cleveland to Shawnee, via Shelby and Newark,...................181   "
Cleveland to New Straitsville, via Columbus.....................201   "
Sandusky to Shawnee, via Newark,................................160   "
Dayton to New Straitsville, via Columbus,.......................134   "
Dayton to Ferrara, via Xenia and Washington, (incompleted
     line)...118   "
Cincinnati to Ferrara, via Muskingum Valley R. R.,.............157   "

4

✳ Toledo is likely to become a chief point of distribution of coal in the region of the lakes. It will have connection with this field by the Atlantic and Lake Erie Railroad, now being constructed from Toledo to Pomeroy, which passes down the center of the Sunday creek valley along the eastern side of the great coal seam. From Ferrara, a mile below the Sands bank, where the seam is found in maximum development, to Toledo, by this line is 174 miles. The grades are light, *north* of the coal-field; but in the field are two tunnels, at which the grades are 52 and 66 feet respectively. Coal trains would, doubtless, be made up *north* of these tunnels, say ten miles from Ferrara; as the trains from Straitsville on the Hocking Valley Railroad are made up near Logan, for a like reason.

✳ The Columbus and Ferrara Railroad will connect with the Atlantic and Lake Erie near Pleasantville; the two companies completing and owning the line in common from the junction to Ferrara. The maximum grade on this line from the upper tunnel, ten miles north of Ferrara to Columbus, is, I am reliably informed, 26 feet to the mile. The distance from Ferrara to Columbus, via this line is 59 miles. These two roads will give the Sunday creek valley coal excellent outlets, by short lines and easy grades, to the great markets and distributing points of the West and North-west.

The Cincinnati and Muskingum Valley Railroad Company, —now under the control of the Pennsylvania Company—is about to construct a short branch from the neighborhood of McLuney down Rogers' fork and West fork of Sunday creek, a distance of about twelve miles; and also, to make arrangements with the Atlantic and Lake Erie Railroad for the use of its line from New Lexington to Ferrara. By this line the distance from the center of the Sunday creek coal-field to the convenient dumpage ground of the Little Miami Railroad at the bridge in Cincinnati, will be but 156 miles. All things considered, I think it will be conceded that this will be one of the very best lines of coal supply for Cincinnati yet completed or projected.

The Pennsylvania Company are now securing a direct and

favorable route, from Sunday creek to Cleveland by constructing a short line of about 20 miles between Dresden—the
present terminus of the Cincinnati and Muskingum Valley
Railroad—and **Oxford**, on the Cleveland, Mt. Vernon and
Columbus **Railroad, both of** which railroads **are** under the
control of the Pennsylvania Company. **The distance from the**
center of the Sunday creek coal-field **to Cleveland will be 166
miles.** As this will be at **least 20 miles shorter than the
shortest** of the other lines, **and with a more favorable grade,
it is safe** to say that it will **be the chief channel of that vast**
coal and iron trade which **Cleveland and this mineral district**
will reciprocally create and command.

By constructing another link from a point on the Cleveland,
Mt. Vernon and Columbus Railroad to Loudonville, on the
Pittsburg and Fort Wayne Railroad, direct connections may
be made with Toledo and the Northwest which will compare
in distance favorably with most other roads.

By these lines of railroads penetrating the upper Sunday
creek valley Columbus, Cincinnati, Toledo and Cleveland,
which are, or will soon become, the great centers of consumption and distribution of Ohio coals, will have most favorable
connections, considering both distances and grades, **with** the
best coal of the Hocking valley and its **tributaries.**

But we must not lose sight of the fact that all the lines of
the roads now or hereafter constructed into this coal field will
soon be connected by short independent lines, or extensions,
so as to give every part of the field direct communication with
each of the great railroads constructed for the **shipment of
its coal.**

I have here spoken only of **the railroads** that are **or are**
being constructed. Many **others** are **proposed,** and several
will, doubtless, soon **be built,** *e. g.*, an extension **of** the
Marietta and Cincinnati road up Snow fork or Sunday creek ;
the McConnellsville and Ferrara Railroad, and the road
from Lancaster via Shawnee to Sunday creek, which last
named road is shown by the survey now being made to be a
most direct and favorable route, and which, if extended

westward as proposed, will give to Springfield, Dayton and the country beyond, a much needed connection with this coal-field.

RECAPITULATION.

We find the *general situation* of the coal properties of these companies most advantageous, as located upon the western margin of the great Allegheny coal-field and proximate to the great coalless district beyond. These lands are so centrally located, that the coal will radiate in almost all directions—to the Lakes on the north and northwest, to Cincinnati on the southwest, and to the vast intermediate area.

The chief seam of coal on these properties is of unusual *thickness*, as has been shown by the many sections given in this report. The *quality* of the coal is also of peculiar excellence, and is adapted to the manufacture of iron to all metallurgic operations, to gas making, to the generation of steam, and to household use. This is shown by the very large number of careful analyses of the coal made by Prof. Wormley, whose skill and accuracy as a chemist, are universally conceded. The coal is also well situated for *easy and profitable mining*.

The lines of railroads, now built or in process of construction, are such that nothing more can be desired for the *successful distribution* of the coal to the great markets. In all respects therefore, these properties are most desirable. At no very distant day the supplies of coal in Europe will begin to fail, and the United States will become the great manufacturing nation of the civilized world. These lands, very valuable now for immediate use, must grow more and more valuable in the future. Their area is so large that there is scarcely a possibility of exhausting their mineral resources for several generations to come.

Very Respectfully Yours,

E. B. ANDREWS.

COLUMBUS, OHIO, December, 1872.

YORK

JACKSON

PINE

BEARFIELD

PLEASANT

MONDAY CREEK

FLICK

MONROE

GORE

WARD

TRIMBLE

GREEN

STARR

YORK

DOVER

WATERLOO

ATHENS

DEERFIELD

UNION

HOMER

AMES

CANAAN

MAP
of the
GREEN FIELD
Coal Region.

Geo. S. Lester, C.E.

EXPLANATIONS.

Rail Roads in operation

www.ingramcontent.com/pod-product-compliance
Lightning Source LLC
Chambersburg PA
CBHW022028190326
41519CB00010B/1631